Seeing Statistics

User's Guide

www.duxbury.com

Quality and innovation

Seeing Statistics^{SM}
User's Guide

Gary McClelland

University of Colorado, Boulder

with documentation by Charles Seiter

Duxbury
Thomson Learning™

Pacific Grove • Albany • Belmont • Boston • Cincinnati • Johannesburg • London • Madrid
Melbourne • Mexico City • New York • Scottsdale • Singapore • Tokyo • Toronto

Sponsoring Editor: *Curt Hinrichs*
Marketing Team: *Tom Ziolkowski*
Media Editor: *Marlene Thom*
Editorial Assistant: *Carrie Izant and Sarah Kaminskis*
Production Editor: *Kirk Bomont*
Manuscript Editor: *Margaret Pinette*
Permissions Editor: *Mary Kay Hancharick*
Cover Design: *Denise Davidson*
Print Buyer: *Vena Dyer*
Typesetting: *Forbes Mill Press*
Printing and Binding: *Webcom Limited*

About the Author

Gary McClelland started studying statistics in high school after a scientist commented on Gary's science fair project, "You have some pretty good research ideas, but it would help if you knew more about statistics." Thus began Gary's life-long love of probability, statistics, and the general problems of making decisions under uncertainty. After graduating from the University of Kansas in 1969 with majors in psychology and mathematics, he spent a summer collecting data on the islands of Micronesia for a joint psychology/anthropology project studying modernization and rapid acculturation. His graduate study at the University of Michigan concluded with an M.A. in statistics (1972) and a Ph.D. in mathematical psychology (1974). Immediately thereafter he joined the faculty at the University of Colorado, where he is now a professor of psychology, and has remained there and will continue to remain there for his entire academic career. He has spent all his adult life at major public universities (Kansas, Michigan, and Colorado) and loves everything, except for committee meetings, about the life at those institutions—the research, the teaching, the diversity of students and colleagues, the intercollegiate athletics, the cultural events, the controversies, and the intellectual excitement.

For over 25 years at Colorado, Gary has taught statistics and other quantitative courses such as measurement and scaling. He is known among his colleagues for his innovative teaching methods and for the integration of technology—first calculators, then computers, and now the Internet—into his classroom. Gary believes the true time to celebrate New Year's is when the new students arrive in the fall. He enjoys each day being in the classroom or lab with students. In addition to teaching the technical details of statistics, Gary enthusiastically communicates that the ability to characterize uncertainty and to make principled decisions in face of such uncertainty is one of the great intellectual achievements of the millennia. Besides teaching statistics, Gary has also published a number of journal articles and book chapters on statistical methods and measurement and scaling and has co-authored a graduate statistics textbook. He frequently consults on statistical issues and has testified about such issues in a number of legal hearings and trials.

Gary's research focus is judgment and decision making, especially those judgments and decisions made by individuals or families that in the aggregate have important social consequences. He has studied such issues as how preferences for having families of a particular sex composition influence decisions about having more children; how concerns about safety, convenience, and efficacy are traded off in contraceptive decisions; and how home owners respond to information about radon risk levels. Other research topics include risk judgments and risk communication, and measuring difficult-to-measure economic values for environmental quality. His most recent research concerns using interactive Web graphics to aid decision making and the study of how decisions are made on the Web. He is a founding member of the Society for Judgment and Decision Making and a charter fellow of the American Psychological Society. He has published numerous scientific journal articles and several books based on research funded by National Institute of Child Health and Human Development, Office of Naval Research, U.S. Environmental Protection Agency, U.S. Forest Service, National Science Foundation, Russell Sage Foundation, General Motors Research Laboratories, and other agencies and companies. Gary's constant use of statistics and quantitative methods in both basic and applied research have sharpened his thoughts about what is important to understand about statistical methods.

Gary lives in the foothills of Boulder, Colorado, with his wife, Lou, who was his high school debate partner and who is now the university's institutional data maven. When not critiquing each other's data analyses, they enjoy biking and wildflowering together and following the adventures of their daughter Abby as she pursues a legal career. Gary is also an avid participant in the Colorado lifestyle including running, road and mountain biking, rock and mountain climbing, drinking lattés, and being outside as much as possible. Gary is also a francophile; he hopes to revise Seeing Statistics from a rental house in Provence if he can find one with a high-speed Internet connection. Gary reads and responds to as much email as he can. Write to him at gary.mcclelland@colorado.edu, and find out more about him at his Web site at http://psych.colorado.edu/~mcclella/.

Contents

Preface

Seeing Statistics is a new approach to teaching statistics using the World Wide Web. *Seeing Statistics* would not have been possible if limited to the printed page. Instead, this webbook uses capabilities of Web browsers to present traditional statistical concepts visually in a way that was never possible before. *Seeing Statistics*'s three guiding design principles, discussed in detail below, are:

- Visual: The important principles of statistics are remarkably easy when they are presented visually.
- Active: Involvement of the student greatly facilitates learning.
- Engaging: An active, visual approach to statistics is fun.

In other words, the *Seeing Statistics* approach is to see it, do it, and enjoy it!

Visual

The Web allows us to move images around easily and cheaply. In printed textbooks and in classroom lectures, graphs, charts, and images are expensive. Publishers never can afford to allow authors to include all the graphs and charts they want. Teachers can never draw enough pictures on the board or display enough transparencies to satisfy student needs for visual representations. The Web solves those problems.

Images are remarkably powerful educational tools. Almost every statistics textbook in its chapter on probability has a few images of probabilistic devices such as a die. However, unlike in any printed textbook, the die in *Seeing Statistics* tosses when the reader clicks on it. After a few clicks, you notice that the die lands with different faces showing on top. Then almost everyone starts wondering: Is this random, or is it a fair die? At this point a helpful instructor can raise the question: What do you mean by "random" or "fair"? And how could we collect and record observations of the die to decide if it is random or fair? In short, this simple interactive image leads to the fundamental issues in statistical inference in a way that words can never do.

Students can experiment extensively with virtual dice.

Statistics is inherently a geometric discipline. It is impossible to imagine that statistics could have developed before geometry. While we statistics teachers *tell* our students about geometric concepts such as "sums of squares," we almost never *show* the students the actual squares. *Seeing Statistics* changes all that. You will see the squares, lines, and other geometrical concepts that underlie statistics. More importantly, you will see those squares and lines change as you *do* statistics visually.

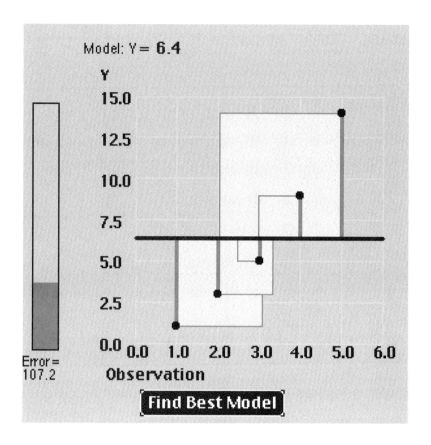

An applet measures squared errors to a center line.

Tables of statistical values are an important part of statistics. But long textbook explanations of how to use those tables are not much fun. And everyone gets confused about how the tails in the table match up with the desired probabilities. The visual "tables" in *Seeing Statistics* provide not only the necessary numbers but, more importantly, a visual representation of the density curves and tail areas.

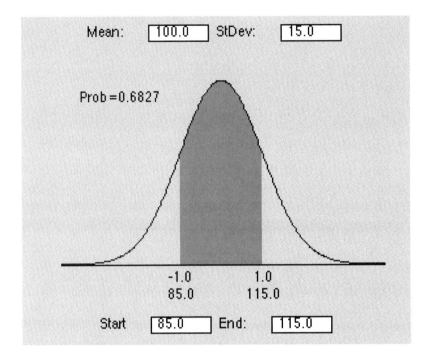

Another applet measures areas under a normal curve.

An important technical note is in order. Some experienced Web users have learned that Web pages with lots of large images can turn using the Web into the World Wide Wait. *Seeing Statistics* avoids this problem by downloading small Java™ applets to your computer. These applets, generally smaller than a single small Web image, quickly generate thousands, or in some cases millions, of images on your computer as they are needed. Also, you are in control of the images. Rather than being forced to download and see the images I want you to see, you are in control so that you can see the images that best help you understand statistics.

Active

Research from cognitive psychology demonstrates that active learning leads to far better understanding than does passive consumption of information. Listening to lectures, watching educational videotapes, and reading textbooks can all too easily become passive, "lean-back" learning. *Seeing Statistics* actively involves the student on each page to promote active, "lean-forward" learning. Rather than watching video demonstrations, students interact directly with the graphics to provide their own demonstrations, demonstrations that they control. When I was a beginning teacher I sometimes gave those lectures that magically transferred the notes from my notebook to the students' notebooks without going through either my or their heads. That can't happen with *Seeing Statistics*.

In *Seeing Statistics* the student controls the pace. Students who may benefit from extra time exploring an interactive graph are not pushed along to the next topic by a fixed pace. Students who have mastered a concept in an interactive graph can move on before they become bored. There are also Discovery activities for each interactive graph that help students learn the most about each concept. Although the interactive graphs are designed to communicate basic statistical principles, they also embody advanced principles. Continued active work with each graph is rewarded by enhanced understanding.

The activity in *Seeing Statistics* is not a lot of drudgery calculations. Instead, the emphasis is on concepts with calculations to be done by statistical software packages or spreadsheet programs.

Engaging

There is no reason that statistics cannot be fun! While the interactive graphs stop short of being arcade video games, they are engaging and fun. Just being in control makes *Seeing Statistics* more fun than being controlled. Interacting with the graphs often reveals surprises, intriguing patterns, and "ah-ha" experiences.

Real data are used throughout *Seeing Statistics* to answer real substantive questions. And if you do not like the example on the main text page, a single click substitutes examples from four different application areas: (a) psychology, education, and social sciences; (b) business, management, and economics; (c) engineering and physical sciences; and (d) biology, medicine, and public health. And if those are not enough, you may enter your own data into almost any of the interactive graphs. One click away for each example are results from four different computer statistical programs: StatView, JMP, Minitab, and Microsoft® Excel.

Easy navigation and many one-click pedagogical aids, including glossary, references, graph help, discovery questions, equations, check-up questions, and historical notes, make *Seeing Statistics* very user friendly. Each chapter begins with an overview and a section titled "Why Am I Learning This?" and concludes with a review and exercises.

Start Seeing Statistics!

These are far too many words about *Seeing Statistics*. The best way to understand what *Seeing Statistics* is about, to experience the educational power of interactive images, to gain from active learning, and to have fun is to get started! Load up *Seeing Statistics* in your favorite browser, and start seeing, doing, and enjoying statistics.

Acknowledgments

There are so many people to thank for their help, encouragement, and sympathetic understanding for this novel project in statistical education. Special thanks go to Jon Roberts, who first showed me how a Java™ applet could help us see statistics. Many kind people at the University of Colorado provided technical help, statistical advice, and moral encouragement, including Richard Cook, Charles Judd, Ernie Mross, Peter Polson, David Rea, Taki Maghjee, Mary Ann Shea, Jim Rebman, Michael Lightner, Marty Goldman, Bobby Schnabel, Donnie Lichtenstein, and Bill Oliver. I am indebted to the members of the Boulder Java™ Users Group who generously teach others about Java™ just because it is fun. Hundreds of students over the years have granted me the privilege of being their teacher and they in turn taught me so much about what works in teaching. Former students like Mina Johnson, who kept asking "Why can't we have more pictures?," motivated this visual approach to statistics.

Seeing Statistics benefited from helpful critiques by a number of statistics educators, including Subra Chakaborti, University of North Carolina, Chapel Hill; David Howell, University of Vermont; John Dutton, North Carolina State University; Tom Gatliffe; Adele Cutler, Utah State University; Ruth Maurer, Walden University; Robert Miller, University of Wisconsin, Madison; and Antonie Stam, University of Georgia. Their detailed comments offered the sympathetic encouragement necessary for this project to continue while at the same time providing the sharp, but friendly, criticism necessary to shape and to refine *Seeing Statistics*. They of course are not to blame for any errors that remain or for any hardheaded insistence on my part to do it a particular way.

Invaluable assistance was provided by several different units of Thomson Learning. Of special note are Bill McLaughlin of Thomson.com for help on server issues and Tracey Claude, Joe Hoover, and Gary Ollinger of the Thomson Media Group for help with design, graphics, colors, scripting, layout, and template construction. Anyone who saw my early versions knows that those individuals made major improvements to the "look and feel" of *Seeing Statistics*.

Unlike the authors of most textbooks, I depended heavily on a variety of computer tools. Of special help were Bare Bones Software BBEdit (www.barebones.com) from Bare Bones Software, MathType (www.mathtype.com) from Design Science, Inc., ImageReady (www.adobe.com) from Adobe®, Mathematica (www.wolfram.com) from Wolfram Research, and Code Warrior (www.metrowerks.com) from Metrowerks. The backup software Retrospect Express (www.dantz.com) from Dantz allowed me to sleep better. None of the interactive graphs would have been possible without Sun® Microsystems' development and support of the Java™ programming language (www.javasoft.com).

Someone asked me what an author's most important consideration should be in selecting a publisher. My advice, which I followed when I signed with Duxbury, is to go with the publisher who has the best people. The wonderful people at Duxbury have exceeded all my expectations. Special thanks are due my editor Curt Hinrichs, who believed in *Seeing Statistics* and a new way of publishing long before anyone else and who has been wonderfully supportive throughout. A number of people at Duxbury and its parent Brooks/Cole have been creative, flexible, tolerant, and encouraging in developing and marketing *Seeing Statistics*. I'm especially grateful to Kevin Connors, Carolyn Crockett, Laura Hubrich, Marlene Thom, and Carrie Izant. I'm grateful to Charles Seiter for developing the print component; Charles understands better than most anyone else what I'm trying to accomplish with *Seeing Statistics*.

Finally, like most authors, I must say that this project could not have been completed without the love and support of my family, who tolerated my long sessions at the computer and who cheerfully came each time I developed a new interactive graphic and called out "Hey, come look at this one!" My daughter Abby and my wife Lou are both statistically sophisticated critical thinkers who provided some of the best criticism and advice.

Gary McClelland

User's Guide

Chapter 1

Map and Contents

Map

The main entry point for *Seeing Statistics* is

<u>www.seeingstatistics.com</u>

Point your Web browser to this URL, and enter the serial number shown on your password insert. Online instructions will guide you in setting up your user name and password, and then you'll see the start page.

At the left side of the page, you'll find the contents icon. Click it, and then you'll see the main map of the site, as listed here for convenience. Each of these topics is itself a link—just click on one and you'll see an expanded list and, next to the contents window, the first screen of that statistics chapter.

0. Introduction
1. Data & Comparisons
2. Seeing Data
3. Describing the Center
4. Describing the Spread
5. Seeing Data (Again)
6. Probability
7. Normal Distribution
8. Inference & Confidence
9. One-Sample Comparisons

Contents

Notes

Chapter 2

Navigation

It's time for a sample tour of *Seeing Statistics*. Because all chapters are structured alike, we'll just examine the features and flow of Chapter 3 as an example.

Preliminaries

A few quick notes about your browsers are in order. First, to navigate in *Seeing Statistics*, start in the Contents window. Identify the chapter you'd like to study, and click that link. Even if you're following *Seeing Statistics* on a nice big monitor (larger than 15"), you may want to hide the browser's navigation toolbar (Figure 2-1) (some browsers do this automatically).

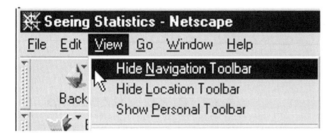

Figure 2-1   Hiding the toolbar

The reason for hiding the browser's navigation toolbar in *Seeing Statistics* is that the little navigation arrows in the upper right hand corner of each *Seeing Statistics* page are the most reliable way to move among the pages. Because of the complex nature of current browser implementation, standard navigation aids such as the Go menu are not necessarily as reliable as *Seeing Statistics*'s own little arrows (Figure 2-2).

Figure 2-2   Navigation arrows

Also, when you step through the very first links in Chapter 0 (Introduction), *Seeing Statistics* will check your browser to see if you have a usable version of the Java™ virtual machine needed for running the program's applets. The applets, small programs that let you interact here graphically with data to "see statistics," are written in the Web-oriented language Java™ and downloaded to your machine. That's why the software checks to make sure you have the right components in your system. Most likely this will not be a problem—browsers from Netscape® and Microsoft® since late 1997 all work properly with the applets.

A Web Tour: Seeing the Center in Chapter 3

So that you can see in print some of the advantages of the *Seeing Statistics* approach, let's step through the pages of Chapter 3. What follows is a large part of what you'll find when you click the Contents icon and then click the link to Chapter 3 in the Contents outline. Be warned, however: *Interacting* with the applets is half the fun and half the "insight value" of *Seeing Statistics*.

Everything you see in the rest of this chapter is what you will find by clicking the little right arrow in the upper right corner of each page of Chapter 3. Alternatively, you can navigate with the Contents window, where all topics are listed as individual Web links, but most users should just step through topics in sequence.

Start (page 3.1)

The story of Chapter 3 really begins at the link for page 3.1. Click this in the Contents window, and after that you can just use the navigation arrows. The story starts with this little preamble:

When working with data, especially when we have a large number of data values, it is very useful to select a representative or typical value for several reasons:

- to describe the data compactly and efficiently,

- to have a value to compare to other data, and

- to be able to identify atypical or unusual values.

Example: Fifteen students in a small honors section of statistics took a 20-item true-false quiz to assess their statistical knowledge on the very first day of class.

If you click on the Survey icon in the right margin, you can take the quiz yourself. Now would be a good time to take this quiz as a pre-test of your statistical knowledge before you go through *Seeing Statistics*.

The scores for the 15 students were:

10 14 9 13 8 12 13 12 7 14 13 14 12 11 11

We would like to find a typical value to represent these data in order to make at least two comparisons:

- to the score we would expect if someone were guessing (i.e., if they knew nothing about statistics) and

- to the typical score obtained on the same test at the end of the semester.

[On the Web site, this first page of the chapter also gives samples from other applied areas, available by clicking on the Applications icon in the right margin (Figure 2-3).]

Help

Discovery

Applications

My Data

Figure 2-3 Right-side icons

In this chapter you will learn to find typical or representative values that will make it much easier to deal with batches of numbers such as those above.

Finding Typical Values

A typical value serves as a compact model representing all the data. Soon we will suggest some principles for finding a good model or typical value, but before we do it is good to have you explore this problem on your own. Our problem is to find a model for these 15 scores from the statistical knowledge quiz:

10 14 9 13 8 12 13 12 7 14 13 14 12 11 11

This graph displays these data. Each value has its own column, and the row value is determined by the score. Thus, the dot in the first column on row 10 represents the first score of 10.

Finding the Center (page 3.2.1)

Just using our eyes, it is often difficult to find the center of the data. Also, it is important to have common definitions of the "center" so that anyone looking at the same data would report the same typical value for the data. For these reasons, we need to have a more formal definition of "center."

One definition of "center": If we have a candidate value for the center or typical value, then we can use the distance from that value (the heavy horizontal line in the graph) to each data value as the amount by which the candidate central value misrepresents each data value.

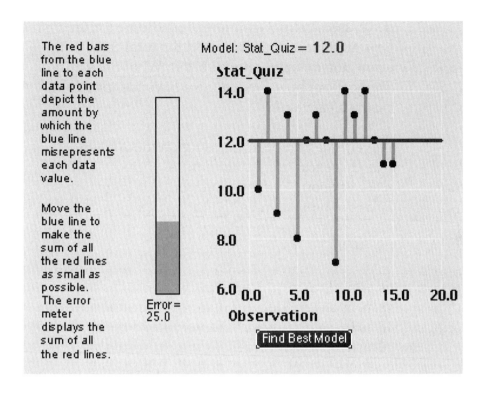

Figure 2-4   Modeling the center

On the Web site, click the Discovery icon.

1. The best model, the best candidate for the typical value, is 12. How much total error is there when we use 12 as the typical value? [Click on the "Find Best Model" button in the graph and read the total error—the total length of all the red lines—under the error meter. (Figure 2-4)]

2. If we used a slightly different candidate for the typical value, say, 12.1, would the total error in the error meter change much? Try it. What does this imply about how precisely we have identified the best typical value?

3. If everyone had been guessing on the 20-item true/false test, we would have expected a typical score of about 10. How much error is there if we use 10 as the typical value? [Click your mouse in the graph on the line for 10, and read the total error from the meter.]

4. Do you think the error when we use 10 as the typical value is so much larger than the error when we use 12 as the typical value that we should reject 10 as a plausible typical value? If so, what would we conclude about the plausibility of the idea that everyone was guessing on the quiz?

Median and Absolute Error

Using the definition of "center" as being the closest to all the data values, the best model or typical value for the statistical knowledge quiz scores is 12. It is easier to see that this typical value is indeed in the center of the scores if we reorder all the scores from smallest to largest.

Original Data: 10 14 9 13 8 12 13 12 7 14 13 14 12 11 11

Reordered Data: 7 8 9 10 11 11 12 12 12 13 13 13 14 14 14

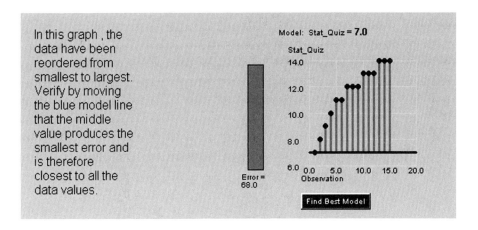

In this graph, the data have been reordered from smallest to largest. Verify by moving the blue model line that the middle value produces the smallest error and is therefore closest to all the data values.

Model: Stat_Quiz = **7.0**

Error = 68.0

Figure 2-5 Modeling the median

The value 12 we identified in the graph as the best model or typical value to represent all the data is indeed the middle data value. In general, the *median* for a data set like that in Figure 2-5 is found this way: Line up the numbers in order, and pick the value in the middle. If there is an even number of values in the set, pick the two values in the middle, take their average, and that will be the median.

The *mean,* in contrast, is the average of all the values in the set. The mean and median behave quite differently with respect to data set outliers (values far away from the others).

The median of the set

$$1 \quad 2 \quad 3 \quad 4 \quad 5$$

is the same as the median of the set

1 2 3 4 1000

The means (averages) of these two sets are wildly different.

Click the check-up icon on the *Seeing Statistics* page to see this little exercise:

A neighborhood committee is concerned about speeding on their residential street, for which the speed limit is supposedly 25 mph. The observed speeds of nine cars during the morning commute are:

23 36 28 42 35 39 29 30 25

What is the median speed of these nine cars?   ___ mph

We now have two candidates to help us identify a typical value or the center of our data. Which is better? Which should we use?

Each has its strengths and weaknesses, so that neither is necessarily better for all situations. In the next pages, we explore the relative strengths and weaknesses of the median and the mean.

Resistance to Extreme Data Values.   Extreme data values do not help us identify the center or typical values. Extreme values are atypical. So we would like our estimate of the center not to be affected too much by extreme data values.

An example from Gould's (1996) eloquent comparison of the median and the mean illustrates their differing resistance to extreme data values. In Gould's example there are five children, one has a penny, one a dime, another a quarter, one a dollar, and the last one ten dollars.

Which estimate of the center, the median or the mean, seems more representative of the amount of money the "typical" child has in this group?

The median is 25 cents (rounded to 30 cents in the graph) and seems fairly typical for all but the largest amount. The kid in this group with ten dollars just isn't typical. The median is not affected by the extreme value of ten dollars. The mean is $2.27 (rounded to $2.30 in the graph) and doesn't seem representative of any of the amounts. The mean is considerably higher (by at least $1.27) than all but the most extreme amount, and it isn't very close to the extreme amount. In short, the mean does not represent any of the money amounts very well.

Lesson: When there are a few extreme observations, the mean may provide a very poor estimate of the center of the data. The median, however, remains in the center of the bulk of the data values and is therefore still a typical value. Thus, the median is more resistant than the mean to extreme values and is therefore preferred for data where there might be a few extreme values, especially on one side of the center.

The median is more resistant to extreme, misleading data values, so it would seem to be the clear choice. However, we also need to consider accuracy. Is the median or the mean more likely to be close to the true value?

To evaluate the relative accuracy of the median and the mean, let's consider how they do when we know the true center of the data. Suppose that the only possible scores are the whole numbers between 0 and 100.

0 1 2 3 4 5 6 7 8 9 10 11 12 13 14 15 16 17 18 19 20
21 22 23 24 25 26 27 28 29 30 31 32 33 34 35 36 37 38 39 40
41 42 43 44 45 46 47 48 49 50 51 52 53 54 55 56 57 58 59 60
61 62 63 64 65 66 67 68 69 70 71 72 73 74 75 76 77 78 79 80
81 82 83 84 85 86 87 88 89 90 91 92 93 94 95 96 97 98 99 100

The center of these 101 numbers, whether we use the median or the mean, is 50. What if we were to select five numbers randomly from this set of 101 and calculate the median and mean of those five numbers? Would the median or the mean be closer to what we know is the true value of 50?

Suppose the five scores we selected randomly were

38 40 50 67 88

In this case, the median is 50, right on the true center, and the mean is 56.6, above the true center. So in this instance the median would be the more accurate estimate of the true center. But we can't be sure whether this one case is itself typical or a quirk. The graph below will allow you to take many different random sets of five scores and determine whether the median or the mean is more accurate.

The applet on page 3.4.3 allows you to try many samples to determine whether the median or the mean (Figure 2-6) is usually the more accurate estimate. The height of the bars indicate the proportion of times the median or the mean, respectively, was the more accurate estimate (i.e., closer to 50).

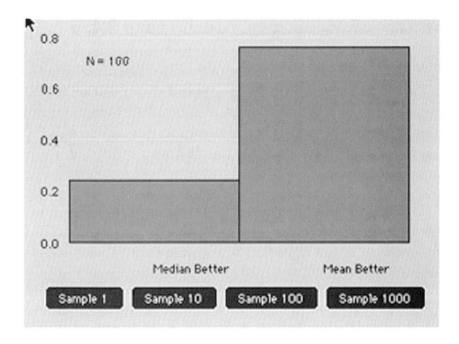

Figure 2-6 Mean vs. median as center

It is not that surprising that the mean is more often closer to the true center than the median. The median can sometimes be inaccurate, because it does not use very much of the information available in the data. While the mean is the average of all the data values, the median is the "average" of just the one or two observations in the very middle of the data. Thus, in situations where there are no extreme observations, the median is often less accurate than the mean.

Summary of Median vs. Mean (page 3.4.3)

So the median is better for describing the center of some data and the mean is better for other data. Statistics is filled with tradeoffs like this. If extreme values are not likely, then the mean is better because it is more efficient—fewer observations will give us an accurate estimate if we use the mean instead of the median. Observations usually cost money, so we prefer the mean.

However, if extreme data values are likely, then the median is better because it is less distorted by extreme data values than is the mean.

At the end of your tour of Chapter 3, you'll find these questions to test yourself (the link is page 3.6).

Newspaper: In your local newspaper, find news articles reporting typical values for data. Do these articles report the median or the mean, or can you tell?

Lottery: The Colorado State Lottery features a number of "scratch-off" games. Consumers buy a $1 ticket; if scratching off areas of the ticket reveals certain patterns, the ticket is a winner. According to the Lottery Commission, approximately 50% of ticket purchases are returned as prizes and the chances of winning for any ticket are 1 in 5. For every 100 lottery tickets sold, about what is the mean amount won by purchasers? What is the median amount won by purchasers? Do you think the mean or the median is more representative of what the typical player of scratch-off games should expect to receive?

Speed: If you were driving on an interstate highway, how could you tell if you were driving faster than, slower than, or the same as the median speed? (HINT: The answer does not involve looking at your speedometer.)

Reading: A school district administers a national reading test to its students. Students scoring below the national median are defined to be "reading below grade level," while students scoring above the national median are defined to be "reading above grade level." The local newspaper (this is a true story, but to avoid embarrassment, the names are omitted) reports that the school board president is "shocked" to learn that 23% of the district's students are reading below grade level.

1. Using those definitions, what proportion of students nationally are reading below grade level?

2. Do you think the school board president should be shocked that 23% of the district's student are reading below grade level?

3. Using those definitions, do you think there are very many students reading at exactly their grade level?

The school board president wants the district to adopt special programs with the goal of reducing the proportion of students reading below grade level to no more than 5%. If that goal is accomplished, what proportion of students will be reading above grade level? Would that make sense?

Notes

Chapter 3

The Applets

Seeing Statistics really has three kind of applets. Some provide animation of models useful for understanding probability—applets demonstrate flipping coins and rolling dice. Others let you interact with standard statistical graphics, so you can move points in a scatterplot, for example, and watch the effect on regression parameters. Finally, some of the applets are the equivalent of statistics software, calculating probabilities directly from distributions.

Here's the whole list, followed by brief descriptions:

AccurateNormal

AccurateNormalQ

BinomCalc

BinomQuantile

Calculator

ChiSq

CircleCompare

CoinFlip

CoinSample

CoinSampleMean

ConfidenceIntervalT

ConfidenceIntervalZ

CorrelationPicture

CorrelationPoints

CountDice

DensityCDF

DiceFlip

DiceSample

DiceSampleMean

GrayMatch

GrayMatch2

GuessQuizCheck

HandleNormal

JitterSlope

LineHow

MeanFit

MeanInfluence

MeanMedianCompare

MedianFit

MedianFit (Audio)

MedianInfluence

MosaicTwoWay

NormalQQPlot

NormalTable

NormOnly

NormZ

PlotHow

PointRemove

PowerNormal

PowerT

RandomQQPlot

RangeRestrict

RegDecomp

RegFit

RegPlain

SampleT

Sampling

ScatHow

SimpleFit

SKQOne

SliderBoth

SliderNormal

SliderStdDev

SlidingNormal

SlidingT

SlopeCalc

StDevSample

StudentCIDemo

StudentT

TTest

ZTest

AccurateNormal

This applet calculates the probability that a given value falls between two values in a normal distribution. Enter the start and end values and press Return, and the applet shows the section of the normal distribution defined by the values, along with the probability (the fraction of the distribution between the two values).

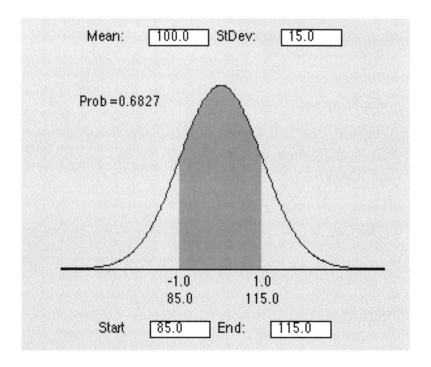

Figure 3-1   AccurateNormal

AccurateNormal is virtually a statistics software kit in itself, because most methods assume that data are normally distributed. For example, if you enter a mean of zero and a standard deviation of one, you have the standard normal curve you use when working with z values. Try these questions, using AccurateNormal to calculate the probabilities:

1. At Wilson Elementary, the mean height of third graders is 135 cm, with a standard deviation of 15 cm. What's the probability that a randomly selected third grader is taller than 155 cm?

2. Rainfall at Cazadero, CA, averages 90 in. a year, with a standard deviation of 20 in. What's the chance of a year with less than 54 in. of rain? Will this likely happen at least once per century?

Note: You may run into problems if you set either Start or End more than four standard deviations away from the mean. At those distances the normal curve is nearly flat with very low values, so there's not much contribution to p (given only to four decimal places). The applet therefore uses an algorithm that expects values to lie between +4 and –4 standard deviations).

AccurateNormalQ

This computes a probability in the same way as AccurateNormal (see above), but it uses a normal distribution defined in terms of z scores instead of "raw data" values.

BinomCalc

Simply enter N (number of trials), p (probability of success on each trial), and start and end values for number of successes, and the applet shows you graphically and numerically the fraction (a probability value) of binomial trials that fall within the range specified by your start and end values. This applet can be used to answer nearly any question—or solve any textbook problem—about outcomes of binomial trials.

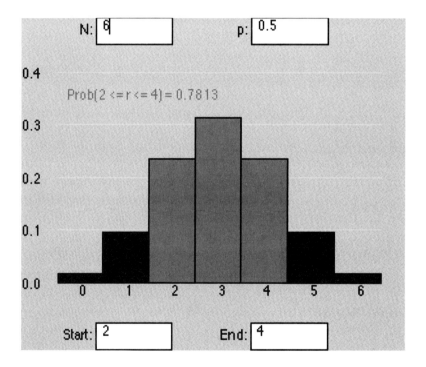

Figure 3-2 BinomCalc

BinomQuantile

For N binomial trials, the outcome "k successes" can range from k = 0 to k = N. This applet takes k as input for a binomial trial experiment and reports the probability of an outcome (number of successes in N trials) falling in the range zero to k.

Calculator

This simple four-function calculator is probably the most familiar applet in *Seeing Statistics,* because its icon appears at the left of every screen on the site. It's just like a standard calculator, except that you push the keys with mouse clicks instead of your finger.

ChiSq

The chi-squared distribution is produced by simulation in direct sampling experiment by this applet. To see a distribution that looks like a smooth chi-squared curve in a textbook, you'll need to use a fairly large number of trials (N = 5,000 or more). The experiment shown in Figure 3-3 gives a result for an experiment with about 2,000 samples, and it's still fairly "lumpy."

CircleCompare

As a perceptual psychology experiment, you are asked to estimate the diameters of circles against different backgrounds.

CoinFlip

This probability experiment lets you flip one and two pennies at a time. Click them to flip again.

CoinSample

This fancier coin-flipping experiment (fancier than CoinFlip, anyway) flips a group of coins and records the total number of heads and tails.

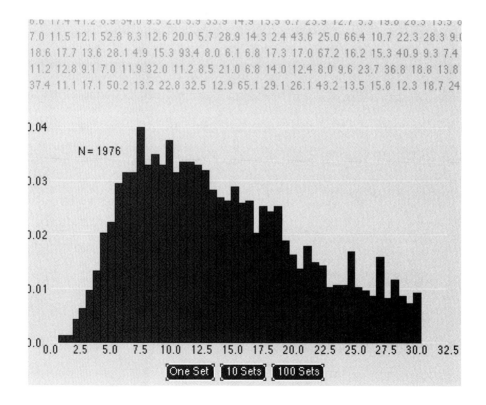

Figure 3-3   ChiSq

CoinSample can solve some of the same problems as BinomCalc. Try these three activities:

1. For a giant family with twelve kids, what's the probability that five of the kids are boys? You can use the $N = 12$ case in CoinSample to model the situation and estimate the probability.

2. Now try using the applet BinomCalc to model the same situation. Again $N = 12$, $p = 0.5$. Unless you've taken several thousand coin samples, the BinomCalc value (an exact value) should show a small but appreciable difference from the value approximated by the CoinSample experiment.

3. Do the same experiments to investigate the situation for $N = 3$. Does the CoinSample estimate of probability more closely approximate the BinomCalc exact value?

CoinSampleMean

This applet records the average outcome (heads and tails are scored as 0s and 1s) of sets of coins flipped in batches. Figure 3–4 shows results for 700 flips of a set of three coins.

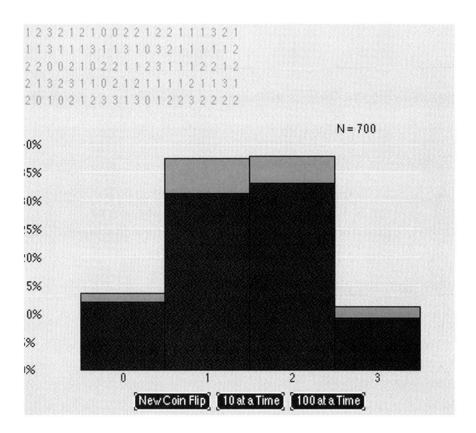

Figure 3-4   CoinSampleMean

ConfidenceIntervalT

By moving the slider at the top of this applet, you can view confidence intervals for the mean of a t-distribution.

ConfidenceIntervalZ

By moving the slider at the top of this applet, you can view different confidence intervals (in terms of z values) around the mean of a normal distribution.

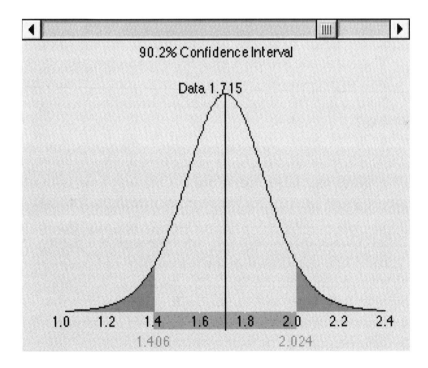

Figure 3-5   ConfidenceIntervalZ

CorrelationPicture

Using the slider, you can call up scatterplots with sets of points showing different values of r (correlation). The applet provides a variety of plot types for characterizing correlation in different contexts.

One of the main uses of this applet is to give yourself some intuition about the implications of different values of r.

1. Move the slider so that r = 0.96. Do the points look strongly correlated to you, compared to the point-cloud in Figure 3-6?

2. Set the slider to produce an r of approximately 0.3. Now click the switch sign button. Switching back and forth between the two sets of points, is it easy to distinguish the directions a fitted straight line might take?

Scroll down the applet window, and you'll find a rich assortment of ways to analyze correlation. One way is to inspect the fit of a regression line: At higher regression values, the data points appear to lie clustered closer to the line. The applet window called Quadrants draws two lines through the data points: a vertical line positioned at the x-mean for the (x,y) points' values and a horizontal line at the y-mean. For a large positive correlation, points will cluster in the upper-right and lower-left quadrants. For small values of r, points are almost equally distributed among the four quadrants.

The Slope "Votes" analysis is reminiscent of the Quadrants method. The applet uses an x-mean and y-mean value to define a center point and displays the slopes of lines from individual data points to this center point. Convex Hull, in contrast, draws the smallest enclosure of the point set. As the hull gets flatter, the points are more highly correlated.

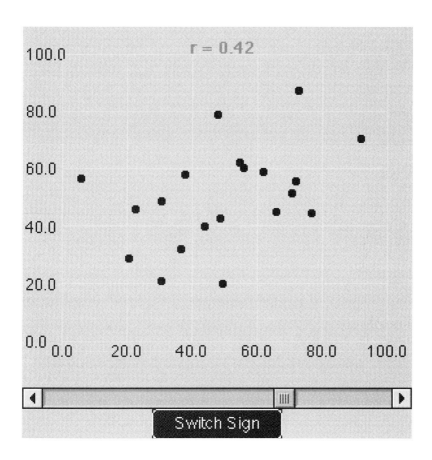

Figure 3-6   CorrelationPicture

The Combined applet window links two of the methods (look closely and decide which methods they are). Perhaps the most exotic correlation applet measures z-score agreement, displaying z-scores of the two-dimensional data sets on a point-by-point basis.

CorrelationPoints

This applet lets you add points interactively to a scatterplot and watch the value of the correlation r change as each point is added.

CountDice

Two six-sided dice produce a set of 36 possible outcomes. Click on any combination of pairs to calculate the probability of the combination.

DensityCDF

The *cumulative* normal reports the probability that a value falls below a certain point in the distribution.

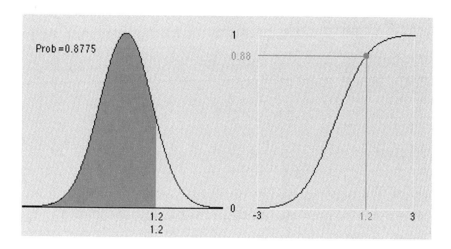

Figure 3-7   DensityCDF

DiceFlip

As evidence that statisticians spend relatively little time in Las
Vegas, this applet for rolling one and two dice is called DiceFlip
instead of DiceRoll.

DiceSample

Roll dice singly or in sets of three or twelve at a time, and record
the results.

DiceSampleMean

This applet records the average number of points showing in a set of dice rolls.

GrayMatch

In this applet, a perceptual psychology experiment, you try to match the shade of gray as a dot against black and white background circles, using a slider to change the shade of gray.

GrayMatch2

This version of the GrayMatch applet records scores for nine attempts to match the two gray values.

GuessQuizCheck

Here's a 20-question quiz, representing the 20-question quiz on statistics in the first chapter. Make guesses yourself, or let the applet guess for you.

HandleNormal

This normal distribution applet lets you determine the fraction of the distribution found within limits set by dragging from either the left or right sides. Update the display by clicking outside this applet window and then back inside again.

JitterSlope

In this plot, you can adjust the line slope by moving the dot at the right end of the line within its limits and look at other slopes by moving the red dot.

LineHow

Use sliders (at the sides of the applet window) to adjust the slope and the intercept of the line.

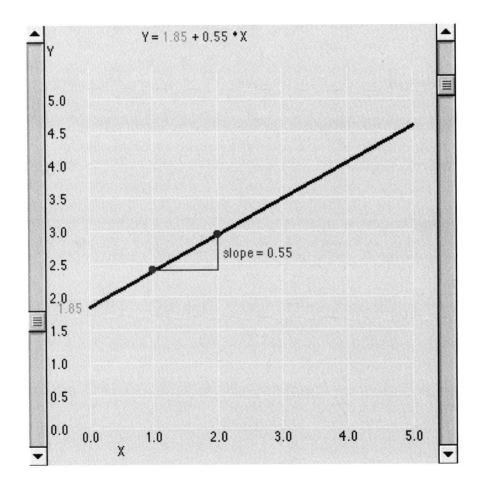

Figure 3-8 LineHow

MeanFit

Guess the mean of a set of points, noticing the set of bars showing distances of the points from your mean value.

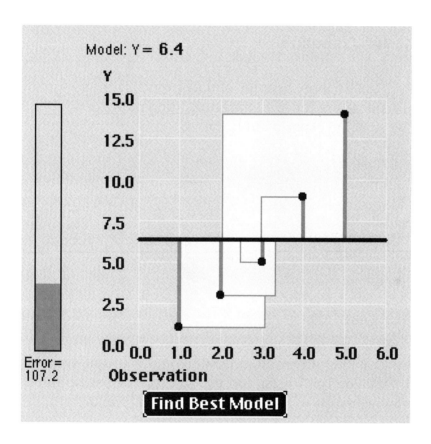

Figure 3-9   MeanFit

MeanInfluence

In this applet, you move a single green point and watch the change in the mean of the whole set of points.

MeanMedianCompare

In sampling from a large number of data points, is the mean or the median a more reliable characterization of the center? This applet runs the experiment for you.

MedianFit

Guess the median value in a set of data points (watch the thermometer at left to observe errors in values).

MedianFit (Audio)

This amazingly loud applet (turn down your computer's audio volume before you run it) also has you guess a median value but reports the results with a tone rather than a graphic.

MedianInfluence

Move the green point in a set, and see the effect on the median value for the set.

MosaicTwoWay

This rather spectacular applet produces mosaic plots, a graphical display type for displaying the statistical results summarized in a chi-square table.

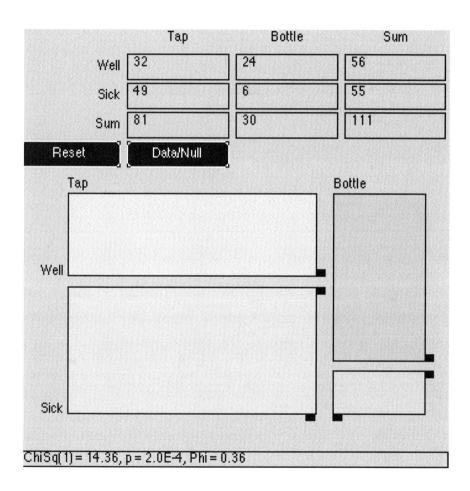

Figure 3-10   MosaicTwoWay

NormalQQPlot

The normal quantile-quantile plot is a visual method for displaying how close a given distribution is to a normal distribution.

NormalTable

Just click on values in the normal-distribution table, and this applet displays the equivalent graphic.

NormOnly

The normal distribution is displayed here, with the abscissa calibrated in terms of z-scores. You can inspect values for one-tailed, two-tailed, cumulative, and "middle-range" sections of the distribution.

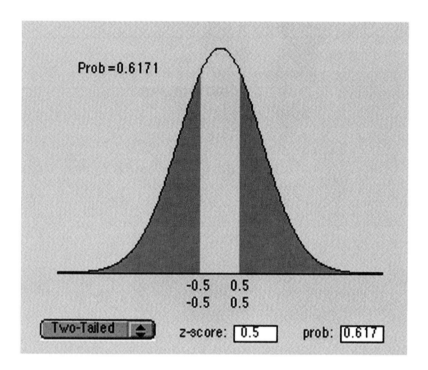

Figure 3-11 NormOnly

NormZ

This applet computes z and a normal curve from input data that you provide. It can be used as a software resource for answering most of the questions in the application topics reported on the *Seeing Statistics* site.

PlotHow

Practice plotting points in a scatterplot. The applet gives you points as coordinate values and checks your plotting.

PointRemove

This applet plots a regression line through a set of points in a scatterplot and lets you remove one point at a time to see the impact on the regression-line fit through the remaining points.

PowerNormal

Observe the power of a hypothesis test (ability to decide correctly between null and alternative hypotheses), using sliders to manipulate normal curves.

PowerT

PowerT performs the same experiment as PowerNormal, for t-distributions instead of normal distributions.

RandomQQPlot

This quantile–quantile plot compares results for a random distribution of values to results for a normal distribution.

RangeRestrict

Subsets of a set of (x,y) values in a plot show different correlation values r. This applet lets you experiment with these subsets.

RegDecomp

Interactively position the regression line through points on a
scatterplot, and this applet shows variation in the average error
for different line positions.

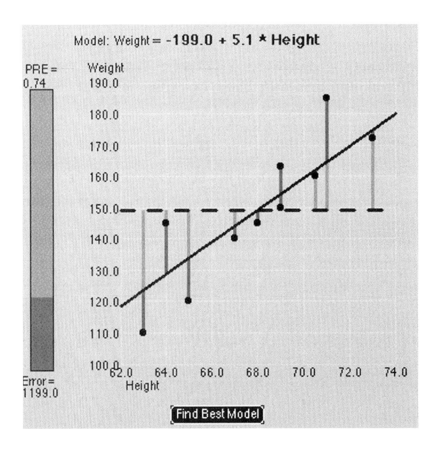

Figure 3-12   RegDecomp

RegFit

In contrast to RegDecomp, RegFit analyzes squares of errors for a regression line position. You will need to experiment with this applet a bit, because it contains essentially the whole story of linear regression for sets of (x,y) pairs.

First, note that the applet displays the squares of distances of points from a line. The object is to minimize the value of the sum of these squares. Second, look closely at two consequences of this object:

- The regression line is essentially pivoted at a point (x,y), where the respective values are the x-mean and the y-mean.

- The slope lets you observe the balance between avoiding producing one big square but also not accumulating too many little squares. What you'll see is that individual outlier points have a big effect on slope in linear regression problems.

- If the fit were simply based on absolute values of distances of points from the line, rather than squares, the slope in many cases would be significantly different and less sensitive to outliers.

RegPlain

True to its name, this plain applet simply reports intercept and slope for a regression line position.

SampleT

Taking points from a normal distribution in sets of five, this applet constructs the N = 5 t-distribution. If you ask it to take thousands of samples, you will see a reasonably smooth distribution.

Historically, William Gosset, writing under the pseudonym "Student," first worked out the form of the t-distribution with an experiment just like this one. The only difference is that, working in the early 1900s with no computer, he had to perform the experiment by manually taking samples from a huge deck of 3,000 specially made cards.

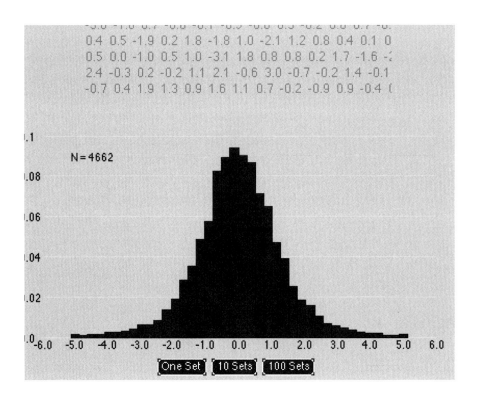

Figure 3-13   SampleT

Sampling

In a grid of 100 blue dots, sample ten at a time, and count the red or green spots underneath to estimate the red/green proportion in the whole set.

ScatHow

Practice making a scatterplot with sets of (x,y) points.

SimpleFit

Move a horizontal line up and down to fit the mean of five points.

SKQOne

This sampling applet makes up a normal distribution with mean 10 as scores of guesses on the 20-question stats quiz.

SliderBoth

Compare two normal distributions using sliders to adjust the mean and standard deviations.

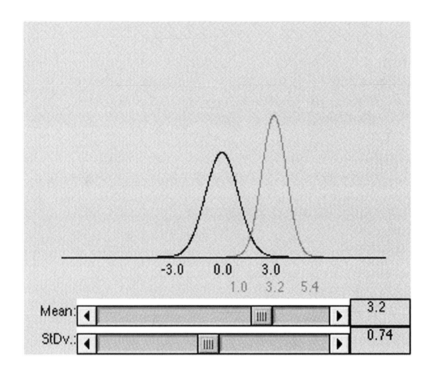

Figure 3-14   SliderBoth

SliderNormal

This applet is half of SliderBoth—it lets you adjust the mean of only one of the distributions.

SliderStdDev

Adjust the standard deviation of one normal distribution in a pair.

SlidingNormal

Observe the range of sample means falling within a 95% confidence interval of the mean of a normal.

SlidingT

This applet is the t-distribution analog of SlidingNormal.

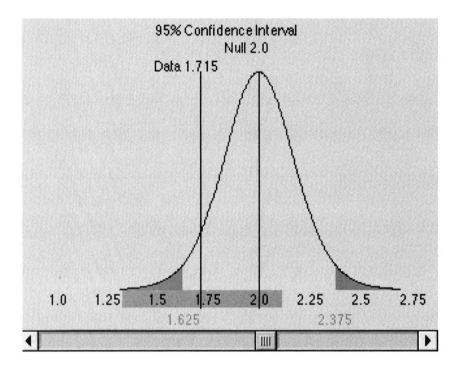

Figure 3-15   SlidingT

SlopeCalc

This applet displays directly a calculation of the squared errors used to determine a regression line slope.

StDevSample

By simulation experiment, observe a sample standard deviation of samples drawn from a distribution of a known population's standard deviation.

StudentCIDemo

Compare shapes and confidence intervals of normal vs. t-distributions.

StudentT

This applet is another interactive demonstration of the properties of t-distributions compared to normal distributions.

TTest

Calculate t from standard deviation, N, sample mean, and hypothesis mean. Results are displayed against the t-distribution for the selected N value.

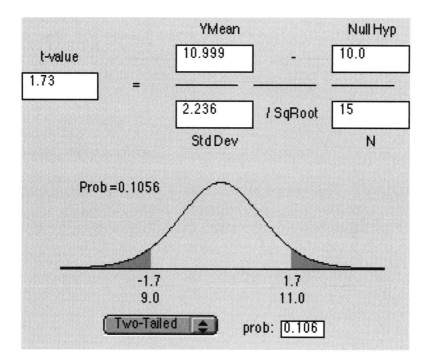

Figure 3-16   TTest

ZTest

Like Ttest, this applet calculates z from basic parameters and displays the result against a normal distribution.

Chapter 4

Applications

Seeing Statistics reflects the belief that you're more likely to understand statistics if it's presented with examples from your own field of interest, rather than showing you only abstract math. Here are examples collected from the *Seeing Statistics* Web site; use these examples as practice data in a statistics program in your course.

Psychology

■ Howell (1995) presents an example about the moon illusion based on the research of Kaufman and Rock (1962). Most people perceive the moon to be larger when it is near the horizon than when it is overhead at its zenith. Kaufman and Rock devised a laboratory apparatus with which to study the moon illusion. Participants made size judgments of an artificial moon near the horizon and overhead. Each person's score was the ratio of the two size judgments. Ratios greater than 1 indicate that the horizon judgment was larger than the overhead judgment; ratios equal to 1 indicate equal judgments; and ratios less than one indicate that the horizon judgment was smaller than the overhead judgment. If their apparatus recreated the moon illusion, typical ratios should be greater than 1. Howell's example reports the ratios for ten participants:

1.73 1.06 2.03 1.40 0.95 1.13 1.41 1.73 1.63 1.56

What is the typical ratio of horizon to overhead size judgments? Is this consistent with the moon illusion?

■ A psychologist evaluates experts in polygraph testing by having a trained assistant answer truthfully to eight questions and untruthfully to eight other questions. Each polygraph expert is scored according to whether the assistant's answers are categorized correctly as being truthful or not. If each polygraph "expert" were simply guessing, we would expect the scores to have a binomial distribution with mean 16(.5) = 8 and standard deviation equal to the SqRoot[16(.5).5] = SqRoot[4] = 2. The mean for a group of ten polygraph experts is 9.1. According to the central limit theorem, the sampling distribution for this mean would be approximately normal with a mean of 8 and a standard deviation equal to 2/SqRoot[10] = .632. What is the probability that the mean for the ten polygraph experts would be 9.1 or greater if they were simply guessing on each answer?

If all the polygraph experts were just guessing which answers were truthful or not, then a mean score on 16 items of 9.1 or greater has a probability of about .04. That is, if they were guessing, a mean of 9.1 or greater would be rare, occurring about 4% of the time. Does this seem a rare enough event to infer that the polygraph experts were doing better than simply guessing?

Business

■ Cryer and Miller (1994, p. 7) report data on the number of photocopies made on an office machine at the University of Wisconsin for 15 consecutive weekdays:

416 556 395 447 238 532 349 390 579 274 621 362 447 440 505

What is the typical number of copies made on this machine?

■   In a particular company, each sales representative is assigned to call on 18 customers each week. Historically, each sales rep takes an order on about eight of those calls or about $8/18 = 44.44\%$. One group of 12 sales reps is given a special training seminar. In the following week, the mean number of sales by reps who have received the training is 8.7. The sales manager wants to know whether this is a higher number of sales than would be expected if performance were no different than it was before the training. If each sales rep after training was no different than a typical sales rep before training, we would expect his or her sales to have a binomial distribution with mean $18(.4444) = 8$ and standard deviation equal to SqRoot[$18(.4444).5556$] = SqRoot[4.444] = 2.108. According to the central limit theorem, the sampling distribution for the mean of 12 sales reps would be approximately normal with a mean of 8 and a standard deviation equal to $2.108/$SqRoot[12] = $.609$.

If the sales reps were no more effective after the training than before, the probability that their mean sales would equal or exceed 8.7 is about .125. That is, about 12.5% of the time their mean sales for a week would be 8.7 or greater. Does this seem a rare enough event to infer that the training program was effective?

Engineering

■   DeVore (1995, p. 21) presents the following example:

The amount of light reflectance by leaves has been used for various purposes, including evaluation of turf color, estimation of nitrogen status, and measurement of biomass. The paper "Leaf Reflectance-Nitrogen-Chlorophyll Relations in Buffelgrass" (Photogrammetric Engineering and Remote Sensing, 1985, pp. 465–466) gave the following observations, obtained using spectrophotogrammetry, on leaf reflectance under specified experimental conditions:

15.2 16.8 12.6 13.2 12.8 13.8 16.3 13.0 12.7 15.8 19.2 12.7 15.6 13.5 12.9

What is a typical value of leaf reflectance under these conditions?

■   In each of ten plants, a company uses 22 sensors in a manufacturing process that exposes the sensors to possible damage. The producer of the sensors claims that 50% of the sensors will still be operational at the end of one year. If the producer's claim were true, we would expect the number of operational sensors at the end of one year to have a binomial distribution with mean 22(.5) = 11 and standard deviation equal to SqRoot[22(.5).5] = SqRoot[5.5] = 2.345. At the end of the year, the mean number of operational sensors across the ten plants is 10.1. According to the central limit theorem, if the producer's claim were true, the mean number of sensors still functioning would have approximately a normal distribution with a mean of 11 and a standard deviation equal to 2.345/SqRoot[10] = .742.

If the producer's claim that approximately 50% would still be operational at the end of the year were true, then the probability would be about .11 that the mean number of operational sensors in the ten plants would be 10.1 or less. Does this seem like a rare enough event to conclude that we should infer that the claim that 50% would be operational is false?

Biology

■ Ott (1993, p. 76) presents this example:

Exercise capacity (in seconds) was determined for each of 11 patients treated for chronic heart failure.

   906 711 684 837 897 1008 1320 1170 1200 1056 883

What is the exercise capacity of a typical patient being treated for chronic heart failure?

■ For each of seven weeks, a biologist grows 25 cultures of the same bacteria and treats them with what is hoped to be an antibiotic. If the cultures had been untreated, it would be expected that about 60% of these cultures would spontaneously die off by the end of one week. If the antibiotic were ineffective, we would expect the number of cultures that die off to have a binomial distribution with mean $25(0.6) = 15$ and standard deviation equal to $\text{SqRoot}[25(.6).4] = \text{SqRoot}[6] = 2.449$. Across the seven weeks, the mean number of cultures that died off equals 16.7. According to the central limit theorem, the sampling distribution of the mean if the antibiotic were ineffective would be approximately normal with a mean of 15 and a standard deviation equal to 2.449/SqRoot[7] = .926.

If the antibiotic were ineffective, the probability would be about .033 that the mean number of cultures dying off across seven weeks would be greater than or equal to 16.7. That is, only about 3.3% of the time would we get a mean of 16.7 or higher if the antibiotic were ineffective. Does this seem rare enough to infer that the antibiotic is effective?

Chapter 5

Software Quick Guides

Seeing Statistics, written in Java™ for the World Wide Web, is currently the most modern application in statistics. Other applications, widely available also in student editions, span a time scale that ranges back to the era of punch cards and building-sized mainframes. So that you can see *Seeing Statistics* in the context of other widely available programs, this chapter shows what common tasks are performed in some examples of commercial statistics software.

Minitab

If you're studying statistics as an undergraduate, it's likely that Minitab is older than you are. But it's been constantly updated, and more textbooks refer to Minitab than any other statistics program. In fact, Minitab offers nearly every statistical test you'll find in a standard textbook, in approximately the same item order as the textbook's chapters. A 30-day free trial version is available for download (Windows only) at:

www.minitab.com

The starting point in Minitab is a table of numbers containing your input data, either typed in directly or imported from a disk file. For generating demonstrations like those in *Seeing Statistics*, Minitab also provides a long list of random-number functions. Here's a set of 500 numbers in a Minitab input table, generated by a standard normal distribution function.

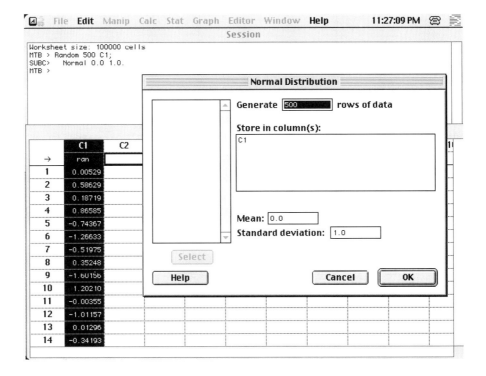

Minitab has histograms as a basic graph type, so it's also possible to check the 500 numbers for approximate normality with a quick graph. As this example shows, and as *Seeing Statistics* explicitly points out, much of classical statistics is based on calculations with normal distributions, and in practice you need 10,000 numbers or more to get a smooth textbook-style normal curve from a histogram.

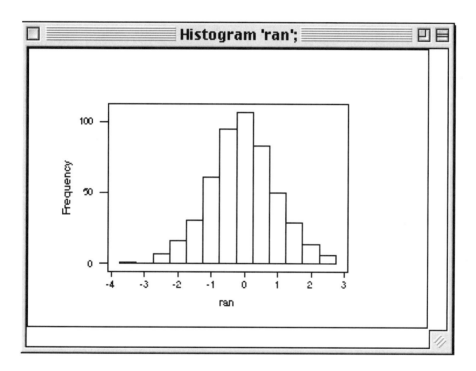

Much of *Seeing Statistics* treats simple problems with Java™
applets that do calculations for t-tests and z-tests. In Minitab,
these two tests appear under the Stat menu as Basic Statistics
options. If you're using Minitab or Student Minitab, just start a
new worksheet table and enter the data from any of the
applications (see below) to compare to the numerical results you
observed online.

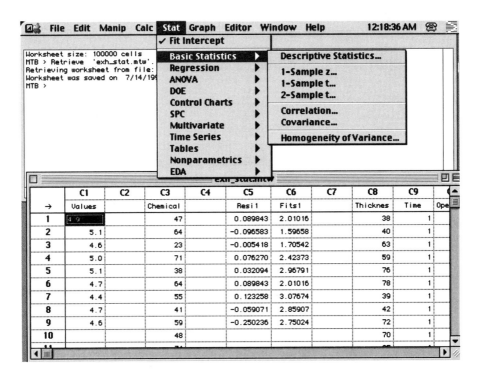

Every version of Minitab provides scores of example data sets for use with its many tests. When you select a z-test or t-test, you'll step through a dialog box in which Minitab explicitly asks you to specify what hypothesis is being tested (test if mean = 0? for example). One of Minitab's customs, which dates to its origins in a computer world before pull-down menus, is that results appear in a window that also logs your commands as text—you can save commands for running the same processes repeatedly on multiple data sets.

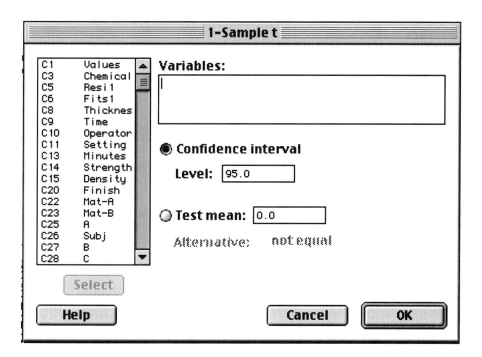

StatView

StatView is a sophisticated stats program, developed in the computer age of graphical interfaces yet oriented toward traditional methods. Its strong point is producing attractive graphical output for publications and reports. In 1998 it was acquired by the SAS® Institute, so it's sold alongside JMP-IN (see below).

Like Minitab, StatView starts with a table of numbers containing your input data, either typed in directly or imported from a disk file. StatView also provides a long list of random-number functions (see figures below). For comparison to other software, we can pop out 500 randoms, generated by a standard normal distribution function.

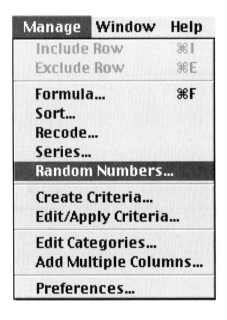

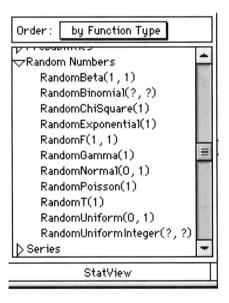

Under the Analyze menu, there's a graphing option that, as you would expect, offers histograms as a basic graph type. Checking the 500 numbers for approximate normality with a quick graph, you may note that StatView fits a normal to the histogram for comparison. Again, you need more than 500 numbers to get a really smooth normal curve.

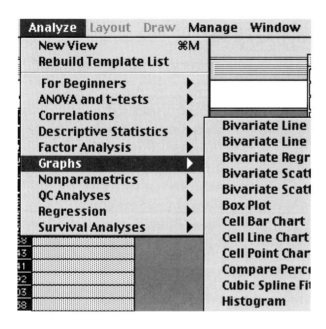

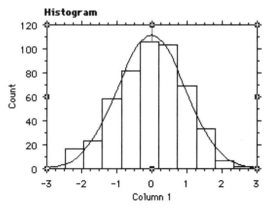

The *Seeing Statistics* Java™ applets that do calculations for t-tests and z-tests find analogs in the first part of the StatView analysis options. To use these to check your results in the Applications pages of the site, just start a new data table and enter the data to compare StatView's answers to the numerical results you observed online.

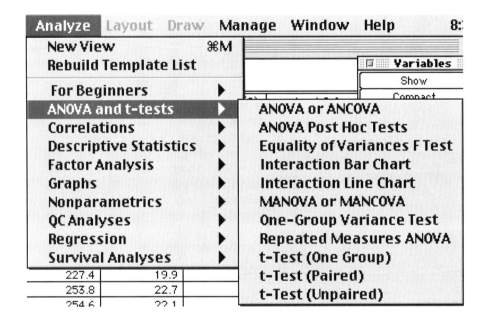

As you'll find if you look through StatView's example lists, the program is oriented toward biostatistics and medical investigations. There are z-tests and t-tests for the simplest cases, packaged along with a huge repertoire of more advanced test types. All the tests are managed through dialog boxes, and StatView provides elaborate step-by-step guidance for the procedures.

JMP IN

JMP IN is the student version of JMP, a statistics program from the SAS® Institute (www.sas.com). It's the most modern software available for students, designed around a comprehensive unified vision of data analysis. Instead of presenting a catalog of tests, as Minitab does, JMP IN gives you simple choices under an Analyze menu that groups many functions under a simple heading. The Analyze menu's "Fit Y by X," for example, covers everything from simple straight-line fitting to complex multivariable schemes beyond the scope of *Seeing Statistics*.

JMP IN also has a standard data table format and a nice calculator function for generating demonstration data (here the data are 500 numbers of a distribution from a standard normal).

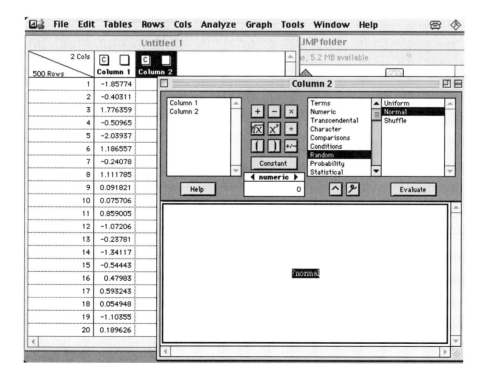

Choosing Analyze (Distribution of Y) produces a histogram (shown here in the horizontal display option) and tables of basic summary statistics.

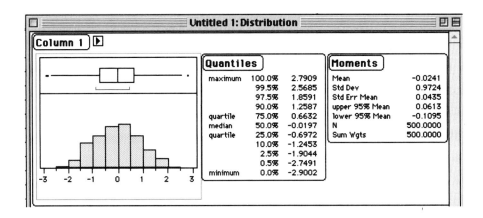

JMP IN takes the position that, in practice, a z-test is just a t-test that uses more data points (t results and z results are numerically the same once N, the number of data points, is 30 or greater). So hypothesis testing for single variables is invoked by clicking the little arrow labeled Gap at the top of the Analyze (Distribution of Y) output, choosing Test Mean = Value. The results graphed here use data from one of JMP IN's own examples.

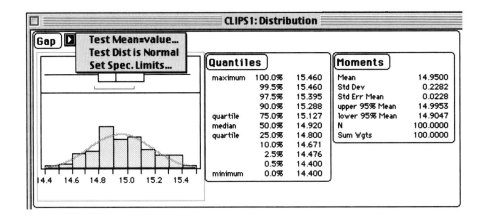

The distinctive approach of JMP IN continues in reporting hypothesis test results. First, notice that the program reports both one-tailed and two-tailed distribution results but in a unique style, reporting Prob > |t| and Prob of values greater and less than t. Also, JMP IN runs the Wilcoxon Signed-Rank test (a common nonparametric equivalent of a t-test) on your data for comparison.

Quantiles		
maximum	100.0%	15.460
	99.5%	15.460
	97.5%	15.395
	90.0%	15.288
quartile	75.0%	15.127
median	50.0%	14.920
quartile	25.0%	14.800
	10.0%	14.671
	2.5%	14.476
	0.5%	14.400
minimum	0.0%	14.400

Moments	
Mean	14.9500
Std Dev	0.2282
Std Err Mean	0.0228
upper 95% Mean	14.9953
lower 95% Mean	14.9047
N	100.0000
Sum Wgts	100.0000

Test Mean=value		
Hypothesized Value	14.97	
Actual Estimate	14.95	
	t Test	Signed-Rank
Test Statistic	-0.876	-289.5
Prob > |t|	0.383	0.315
Prob > t	0.809	0.843
Prob < t	0.191	0.157

Another test available in the Gap pull-down menu is a test for normality. It's the equivalent of the Quantile–Quantile applet used in the later chapters of *Seeing Statistics*. The closer the dots fall to a straight line, the closer the distribution is to normal.

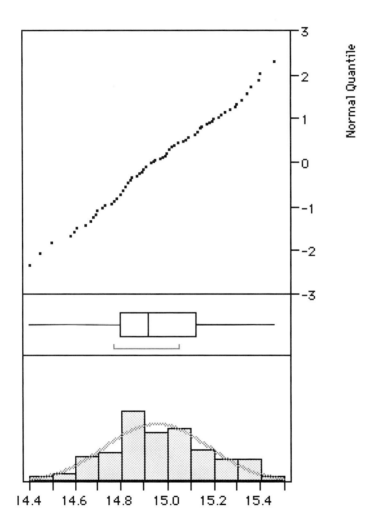

Microsoft® Excel

Microsoft® Excel, part of Microsoft® Office, is a spreadsheet that offers built-in statistical functions and a special add-in called Data Analysis that packages these functions into convenient forms with more "helpful" help than you see for the statistical functions alone.

For example, just as the dedicated statistics programs can generate different sets of random numbers, if you select the Data Analysis choice from Excel's Tools menu (consult Help if you don't have this standard add-in installed), one of the options is Random Number Generation, and a normal distribution is offered. The dialog box for this selection asks you where to put the numbers, and in the case that follows they have just been directed to the first 500 cells in the first column.

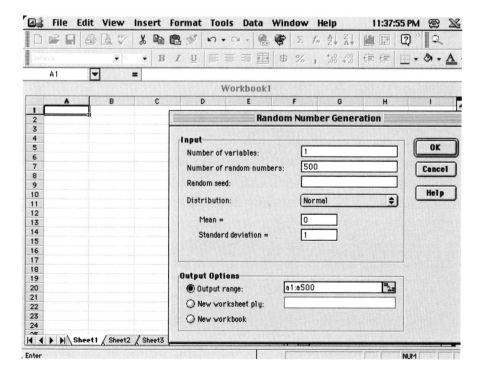

Excel also offers a Histogram choice in Data Analysis—the following figure shows a histogram for the 500 standard normal random numbers. This figure also shows a little floating menu bar for another Excel add-in useful for doing statistics by simulation in the style of *Seeing Statistics*. A demonstration version is available free at www.resample.com.

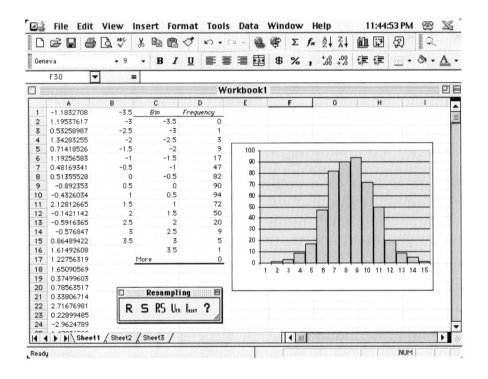

Although the statistical tests in Excel don't approach the scope of Minitab, JMP IN, or StatView, it's still a good software companion for *Seeing Statistics* for two reasons. First, it provides the tests covered by *Seeing Statistics* examples in a relatively simple fashion. Second, because it's a spreadsheet, it's not difficult to construct more advanced statistical tests from the basic tests, with the educational advantage that this approach lets you see the actual calculations involved.

Excel has a standard z-test function for z-tests on a single sample, and the list of tests available in Data Analysis includes two-sample z-tests and three different variations of t-tests. The last figure in this test shows the output from a paired-sample t-test. It's not fancy, but it's complete and easy to compare to the standard test output presented in most textbooks.

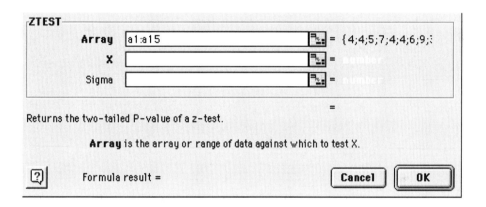

	A	B	C	D	E	F	G
1	4	5		t-Test: Paired Two Sample for Means			
2	4	7					
3	5	4			Variable 1	Variable 2	
4	7	6		Mean	4.53	5.60	
5	4	5		Variance	3.55	3.69	
6	4	7		Observations	15.00	15.00	
7	6	9		Pearson Correlation	-0.29		
8	9	4		Hypothesized Mean Difference	0.00		
9	3	8		df	14.00		
10	4	5		t Stat	-1.35		
11	6	2		P(T<=t) one-tail	0.10		
12	5	4		t Critical one-tail	1.76		
13	2	4		P(T<=t) two-tail	0.20		
14	3	6		t Critical two-tail	2.14		
15	2	8					
16							

Notes

Glossary

absolute error: The difference, without sign, between the actual value and a statistically determined model value.

$$e_i = |Y_i - \hat{Y}_i|$$

alpha: The probability that the statistical test will incorrectly reject the null hypothesis.

analysis of variance (ANOVA): A statistical method for comparing the means of multiple groups.

average: Another name for the *mean*. The sum of all the data values divided by the number of data values.

$$\overline{Y} = \frac{\sum Y_i}{n}$$

beta: The probability of making a *Type II error* when making a statistical hypothesis test. One minus this probability is *power*, the probability of correctly rejecting the null hypothesis.

binomial distribution: The probability distribution describing the number of events (e.g., heads for coin flips) in a given number of trials for a specified probability of the event happening on each trial. The probability distribution describing the number of events (e.g., heads for coin flips) in a given number of trials for a specified probability of the event happening on each trial.

box plot: The box represents the middle 50% of the data values; that is, the data values from the first to third quartiles. The line or dot in the middle of the box denotes the median. Whiskers— lines extending from the box—extend either to the limits of the data or to a maximum length equal to 1.5 times the length of the box. In the latter case, points beyond the lines indicate outliers.

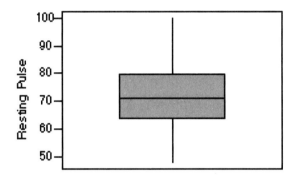

categorical variable: The values of a categorical variable are discrete categories. Usually the categories cannot be ordered from "highest" to "lowest" in any meaningful way. Rather, the categories just define different groups. Gender and religion are examples of categorical variables. Compare *continuous variable*. See also *variable*.

central limit theorem: No matter what the initial probability distribution of individual observations, the sampling distribution of the means of samples of observations from that distribution will increasingly approximate a *normal distribution* as the sample size increases.

chi-square test: A statistical test for determining whether two *categorical variables* are related.

confidence interval: All the values for a parameter that would not be rejected if they were used as the null hypothesis in a statistical test of that parameter. A 95% confidence interval includes all those values that would not be rejected with $p = .05$.

contingency coefficient: Quantifies the degree of relationship between two *categorical variables*.

continuous variable: The potential values of a continuous variable are numbers that can be meaningfully ordered from "highest" to "lowest." Age and gross domestic product are examples of continuous variables. Compare *categorical variable*.

control group: The group in an experiment that receives no treatment. We compare the control group to a treatment group to see if the treatment had an effect, either positive or negative.

correlation coefficient: A measure of the degree to which two variables co-relate. If there is no relation (i.e., if the variables are independent), then the correlation coefficient is zero. See *Pearson correlation coefficient* for a particular example.

covariance: Assesses the degree to which two variables co-vary or vary together. If the two variables are independent then the covariance will equal zero. It is computed as the mean of the products of the mean deviations for each variable.

$$s^2_{XY} = \frac{\sum (X_i - \bar{X})(Y_i - \bar{Y})}{n}$$

cumulative frequency: The total number of observations with values from the minimum up to a certain value.

cumulative probability: The sum of the probabilities from the minimum possible value up to a certain value.

degrees of freedom: The number of independent pieces of information available in the data after various summary statistics have been calculated. For example, if we know the mean for a set of data values, then after seeing $n - 1$ of the data values, we would be able to determine the last one from the mean. For t-test comparisons of two means, there are $n - 2$ degrees of freedom because two summary statistics have been calculated.

dependent variable: A variable observed in an experiment that is not under the direct control of the experimenter. Compare *independent variable*.

discrete variable: A variable that can have only certain discrete values. The same as a *categorical variable.*

dot plot: A simple graph in which each observation is represented by a dot at its score. For example, the scores

7 8 9 10 11 11 12 12 12 13 13 13 14 14 14

produce this dot plot:

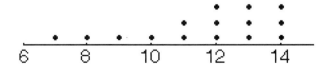

error: The error for an observation is the amount by which a model of the data misrepresents the data value. That is,

$$e_i = Y_i - \hat{Y}_i$$

where $\hat{Y}_i$ is the predicted value from a model such as the *median, mean*, or the prediction from a *regression model*. Error is also used to refer to the total error or all the individual errors added up in some way.

F distribution: A probability distribution for ratios of variances. Most commonly used in *analysis of variance* to compare the ratio of the variance of the means from a number of groups to the expected variance of those means if all the groups were the same.

frequency: The number of times a certain data value occurs in the set of observations.

Gaussian distribution: Another name for the *normal distribution*.

heteroscedasticity: Data are heteroscedastic when the variances within groups of observations are unequal. Data in a regression are heteroscedastic when the variance of the dependent variable depends on the level of the independent variable. Data in a t-test or analysis of variance are heteroscedastic when the variances of the observations within each group are unequal across groups.

histogram: A plot of the frequencies of data values in a set of observations.

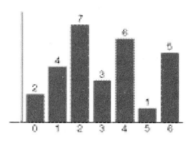

hypothesis test: A test of whether or not the *null hypothesis* should be rejected based on the data.

independent variable: In an experiment, the levels of independent variables are controlled by the experimenter. Independent variables serve as inputs. Compare *dependent variable*.

intercept: In the following equation for a line such as a regression model, the parameter a is the intercept:

$$Y = a + bX$$

The intercept describes the value of Y when $X = 0$. In a graph, it is where the line crosses or "intercepts" the Y-axis. Compare *slope*.

interquartile range (IQR): The difference between the third *quartile* (Q3) and the first quartile (Q1). The IQR indicates the range of scores spanned by the middle half of the data values. The IQR is represented by the box in a *box plot*.

Kruskal-Wallis test: A nonparametric statistical procedure for comparing medians across groups.

lower quartile: The lowest 25% of the scores are equal to or less than the lower quartile score. Often represented as Q1; equal to the 25th *percentile*.

maximum: The largest data value in a set of observations.

mean: The average of the data values, that is, the sum of all the data values divided by the number of data values. The mean makes the *sum of squared errors* as small as possible.

$$\overline{Y} = \frac{\sum Y_i}{n}$$

median: The middle data value in a set of observations. To find the median, reorder the data from smallest to largest and find the middle observation; that is the median. If there are an even number of observations, then there will be two middle values; in that case, the average of those two middle values is the median. The median is also the 50th *percentile*.

median absolute deviation (MAD): The median of all the *absolute errors* or deviations. The MAD describes the spread of the data values away from the model value.

minimum: The lowest data value in a set of observations.

mode: The most frequently occurring value in a set of data.

model: Specifies a predicted or representative value for each observation in a data set. Simple models are the mean or the median for a set of observations. More complex models are, for example, regression lines.

nominal variable: Same as *categorical variable*. Levels of these variables are *noms* or *names.* For example, the names for the levels of the gender variable are "male" and "female."

nonparametric statistics: Statistical methods that make no assumption that the data are from a *normal distribution.*

normal distribution: The probability distribution that describes or approximates the distribution of many variables. Most observations in a normal distribution occur near the mean.

The area under the curve between two values represents the probability that a random observation from the normal distribution would fall between those two values.

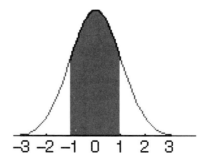

null hypothesis: The hypothesis that nothing is happening in the data that would not be expected by chance. Some null hypotheses are (a) the means of two groups are equal, (b) the slope in a simple regression is zero, (c) the correlation between two variables is zero, and (d) the classification of the data according to one nominal variable is independent of the classification of that data according to a second nominal variable.

observation: A person, place, or thing for which a data value is available or is measured.

one-tailed test: A one-tailed test considers whether the null hypothesis might be wrong in precisely one direction; that is, it allows the possibility that the true parameter might be different from the value of the null hypothesis in only one direction— either only above or only below the value of the null hypothesis. Compare *two-tailed test*.

ordinal variable: A variable whose values indicate only rank and not the distance between values. For example, in a race, an ordinal variable would indicate who was first, second, third, etc., but would not give the time or distance between the racers.

outlier: An extreme or atypical value that is very different from other data values in the set of observations.

*p***-value**: The *p*-value for a particular value of a statistic is the probability of obtaining a value for the statistic that extreme or more extreme if the *null hypothesis* were true. If the *p*-value is less than a specified value (usually .05, but sometimes .01), we reject the null hypothesis.

parameter: An unknown value in a model that is estimated from the data. Examples: mean, median, intercept, slope.

Pearson correlation coefficient: The Pearson correlation coefficient—usually represented by the symbol r—measures the linear relationship between two variables. Values of the correlation coefficient are always between –1 and +1, inclusive. The value $r = 0$ indicates no relationship between the two variables. Positive values of r imply that higher values on one variable are associated with higher values on the other variable. Negative values of r imply that higher values on one variable are associated with lower values on the other variable. The value $r = +1$ indicates a perfect positive linear relationship and $r = –1$ indicates a perfect negative linear relationship between the two variables. Definitional formula in terms of *z-scores*:

$$r = \frac{\sum_{i=1}^{n} z_{X_i} z_{Y_i}}{n}$$

and in terms of *covariance* and *standard deviations*:

$$r = \frac{s_{XY}}{s_X s_Y}$$

percentile: The Pth percentile data value is the score that is equal to or greater than $P\%$ of all the data values. For example, the score equal to the 25th percentile is equal to or greater than one quarter of all the data values (and is also the first *quartile*). The 50th percentile necessarily is the *median*.

population: The set of all possible observations. Compare *sample*.

power: The probability that a statistical test will reject the *null hypothesis* given that some specific alternative model is true. Equivalently, the probability that a *Type II error* will be avoided.

probability: The likelihood that some event will randomly occur.

probability distribution: A description of the probabilities for a number of events will occur or that specific data values will occur.

quartiles: Scores that divide all the scores into four quarters. One-fourth of the data values are equal to or less than the first or *lower quartile* (Q1), one half of the values are equal to or less than the second quartile (Q2, also known as the *median*), and three fourths of all the data values are equal to or less than the third or *upper quartile* (Q3).

R-squared: A coefficient indicating the degree to which one variable predicts another or is related to another. R-squared varies between 0 (no relationship, no prediction) to 1 (perfect prediction).

random sampling: A random selection of possible observations.

range: The difference between the maximum and minimum data values. If the maximum were 90 and the minimum were 10, the range would be $90 - 10 = 80$.

ranks: Data values which indicate only the order of the observations. See *ordinal variable.*

regression coefficients: The parameters in a regression equation. For a simple regression, these are the slope and the intercept.

residual: The difference between a data value and the prediction from a model. See *error.*

sample: A selection of observations from a larger *population* of all possible observations. Ideally, the sample is selected randomly from the population.

sampling distribution: The probability distribution of the statistic one would get by sampling over and over again.

scatterplot, scattergram: A two-variable plot in which each point represents the levels of an observation on each variable. Often the variable on the horizontal axis is to be used to predict the variable on the vertical axis. Scatterplots are useful for viewing *correlations* and for *simple regression.*

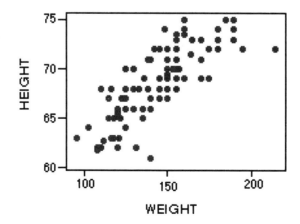

significance level: The pre-specified probability of a Type I error below which the null hypothesis is considered to be so rare as to be implausible.

significant: When the probability of obtaining a statistical value is below the significance level, the statistic is said to be "statistically significant."

simple regression: Fitting a line to the data such that

$$Y = a + bX$$

where *a* is the intercept and *b* is the slope.

skewed: A skewed distribution is not symmetric. That is, it has different shapes on each side of the median. For example:

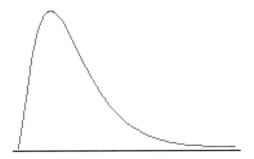

slope: In the following equation for a line such as for a simple regression model, the parameter *b* is the slope.

$$Y = a + bX$$

The slope describes how much *Y* increases (decreases, if its sign is negative) as *X* increases by one unit. Compare *intercept*.

squared error: The *error* for an observation is squared or multiplied by itself. This has the effect of increasing the penalty for large errors.

$$e_i = (Y_i - \hat{Y}_i)^2$$

standard deviation: The square root of the *variance;* describes the typical or "standard" deviation of the data values from the *mean.*

$$S = \sqrt{\frac{\sum(Y_i - \overline{Y})^2}{n-1}}$$

standard error: The *standard deviation* of the *sampling distribution* for any given statistic.

standard error of mean: The *standard deviation* of the sampling distribution of the mean.

standard normal distribution: A *normal distribution* with mean 0 and variance 1.

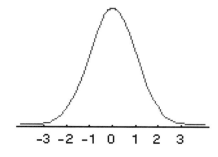

sum of squares: The *squared error* for each observation summed across all observations. Least-squares statistics such as the *mean* and *regression* minimize the sum of squared errors, which equals:

$$\sum_{i=1}^{n} (Y_i - \hat{Y}_i)^2$$

symmetric: A symmetric distribution has the same shape on each side of the median. The normal distribution is an example of a symmetric distribution. Compare *skewed*.

two-tailed test: A two-tailed test considers whether the null hypothesis might be wrong in either direction; that is, it allows the possibility that the true parameter might be either above or below the value of the null hypothesis. Compare *one-tailed test*.

Type I error: Incorrectly rejecting the *null hypothesis* when the null hypothesis is true; a "false alarm." Compare *Type II error*.

Type II error: Failing to reject the *null hypothesis* when the null hypothesis is false; a "miss." Compare *Type I error*.

upper quartile: The 75th *percentile*. The observation that is greater than or equal to 75% of the observations and less than 25% of the observations. Sometimes designated by Q3.

variable: A characteristic, trait, attribute, or measurement that can take on different values. A variable must vary, having at least one value for some observations and another value for other observations.

variance: The typical or average *squared error*. For a population, the *sum of squared errors* is divided by n, the number of observations. More often, interest is in the sample estimate of the variance in which case the sum of squared errors is divided by $n - 1$. That is,

$$Variance = s^2 = \frac{\sum\limits_{i=1}^{n} (Y_i - \bar{Y})^2}{n - 1}$$

z-score formula: The difference between the observation and the mean, divided by the standard deviation. That is,

$$z = \frac{Y - \mu}{\sigma}$$

The *z*-score tells how far an observation is from the mean in terms of standardized units.

References

Agresti, A. (1990). *Categorical data analysis.* New York: Wiley-Interscience.

Cryer, J. D., & Miller, R. B. (1994). *Statistics for business: Data analysis and modeling.* Pacific Grove, CA: Duxbury Press.

Devore, J. L. (1995). *Probability and statistics for engineering and the sciences* (4th ed.). Pacific Grove, CA: Duxbury Press.

Friendly, M. (1992). Mosaic displays for loglinear models. Paper presented at the ASA meetings (Statistical Graphics Section), Boston, August, 1992, Published: *Proceedings of the Statistical Graphics Section,* 1992, 61–68. Available on the Web at http://www.math.yorku.ca/SCS/asa92.html.

Friendly, M. (1994). Mosaic displays for multi-way contingency tables. *Journal of the American Statistical Association, 89,* 190–200.

Gould, Stephen Jay (1996). *Full house: The spread of excellence from Plato to Darwin.* New York: Harmony Books.

Hamilton, L. (1996). *Data analysis for social scientists.* Pacific Grove, CA: Duxbury Press.

Howell, D. C. (1995). *Fundamental statistics for the behavioral sciences* (3rd ed.). Pacific Grove, CA: Duxbury Press.

Howell, D. C. (1999). *Fundamental statistics for the behavioral sciences* (4th ed.). Pacific Grove, CA: Duxbury Press.

Johnson, R., & Kuby, P. (1999). *Just the essentials of elementary statistics* (2nd ed.). Pacific Grove, CA: Duxbury Press.

Judd, C. M., & McClelland, G. H. (1989). *Data analysis: A model comparison approach.* San Diego, CA: Harcourt Brace Jovanovich.

Kaufman,L., & Rock, I. (1962). The moon illusion, I. *Science, 136,* 953–961.

Keller, G., & Warrack, B. (1997). *Statistics for management and economics* (4th ed.). Pacific Grove, CA: Duxbury Press.

Mann, J. I., Vessey, M. P., Thorogood, M., & Doll, R. (1975). Myocardial infarction in young women with special reference to oral contraceptive practice. *British Medical Journal, 2,* 241–245.

Ott, R. L. (1993). *An introduction to statistical methods and data analysis.* Pacific Grove, CA: Duxbury Press.

Rosner, B. (1995). *Fundamentals of biostatistics* (4th ed.). Pacific Grove, CA: Duxbury Press.

Townsend, T. R., Shapiro, M., Rosner, B., & Kass, E. H. (1979). Use of antimicrobial drugs in general hospitals I. Description of population and definition of methods. *Journal of Infectious Diseases, 139,* 688–697.

Tufte, E. R. (1997). *Visual explanations: Images and quantities, evidence and narrative.* Cheshire, CT: Graphics Press.

Index